W O R L D   O F   W O N D E R

Published by Creative Education
123 South Broad Street
Mankato, Minnesota 56001

Creative Education is an imprint of
The Creative Company.

Art direction by Rita Marshall
Design by The Design Lab

Photographs by Peter Arnold (John Cancalosi), Robert
E. Barber, The Image Finders (Kenny Bahr, Rob
Curtis/The Early Birders, Bill Leaman, Michael
Lustbader), Innerspace Visions (Mark V. Erdmann), JLM
Visuals (Don Blegen, J. C. Cokendolpher, Richard P.
Jacobs, Breck P. Kent, Marypat Zitzer), Tom Myers,
Root Resources (Steve Herrick, Robert W. Knupp, Dave
Tollerton), James P. Rowan, Randal D. Sanders, Tom
Stack & Associates (Jeff Foott, Thomas Kitchin, Kitchin &
Hurst, Mark Newman, Brian Parker, Milton Rand, Mike
Severns), Elwin Trump, Visuals Unlimited (Jeff J. Daly)

Library of Congress Cataloging-in-Publication Data

Hoff, Mary King.
Mimicry and camouflage / by Mary Hoff.
p. cm. – (World of wonder)
Summary: Discusses how various animals use mimicry
and camouflage to protect themselves or to lure prey
to them.
ISBN 1-58341-237-9
1. Mimicry (Biology)–Juvenile literature. 2. Camouflage
(Biology)–Juvenile literature. [1. Mimicry (Biology) 2.
Camouflage (Biology)] I. Title.

QH546 .H62 2002
578.4'7–dc21   2001047884

First Edition

9 8 7 6 5 4 3 2 1

cover & page 1: an io moth
page 2: an orchid mantis
page 3: a sea snail on coral

Creative Education presents

W🌍RLD OF W🌍NDER

MIMICRY AND CAMOUFLAGE

BY MARY HOFF

Insects that look like flowers 🕷 Flowers that look like insects ☀ Plants that smell like dead meat 🦃 Birds that sound like snakes 🐐 The world is full of creatures that are not what they seem to be. Some **mimic** other living things. Others are **camouflaged**, which means they blend in with their surroundings.

THE SURPRISES AND DISGUISES involved in mimicry and camouflage are all around us. In the forest, at the bottom of the sea, and in our own backyards, living things are constantly tricking each other. These tricks are not for fun, though. They are **adaptations**: traits that help plants and animals to survive.

Chameleons are known for changing colors

HIDDEN CRITTERS

In the corner of a yard is a pile of dead leaves. Suddenly, a bit of the pile hops away. What had seemed to be a dead leaf was really a blotchy brown toad. The toad's color and markings disguised it in a way that made it blend perfectly into its surroundings.

Camouflage, also known as cryptic coloration, helps some living things avoid becoming another creature's meal. Many birds' eggs have speckles and spots that make them harder for a predator to see than a bright white egg would be. The dapples on the back of a white-tailed deer fawn allow it to blend into the pattern formed by sunlight filtering through tree leaves in the forest.

Of course, **predators** can play the hiding game, too. The stripes of a tiger help it blend into tall grass so it can sneak up on animals such as wild pigs and

NATURE NOTE: *A zebra's stripes help disguise it by making it hard for predators to tell where the zebra's body stops and its surroundings begin.*

There is a frog hidden in this pile of leaves

NATURE NOTE: *The cuckoo lays eggs in other birds' nests and leaves them to be cared for by the hosts. The cuckoo's eggs resemble the eggs of the host.*

White fur helps polar bears blend into snow

antelope. If a polar bear were black instead of white, it would have a much harder time surprising seals and other **prey** animals amid snow and ice.

STICKS AND STONES

Instead of coloring that helps them blend into the environment, some animals have disguises that make them look like other living or nonliving

NATURE NOTE: Phasmid, *the scientific name for sticklike and leaflike insects, is related to the words "fantasy" and "fantastic."*

This "twig" is really a caterpillar in disguise

Six of these "rocks" are actually small plants

things. Some katydids (a kind of grasshopper) look almost identical to leaves. Insects known as phasmids look like sticks or leaves.

❈ The pebble plant, which grows in southern Africa, escapes the attention of ostriches and other animals that might like to eat it because it looks like stones scattered on the desert floor. Some species of spiders and moths escape enemies because they look like bird droppings. Some leaf beetles look like caterpillar droppings.

Some caterpillars look like bird droppings

LOOKS LIKE TROUBLE

Some kinds of mimicry make an animal more obvious instead of less obvious. How does this help it to survive? By sending a message to predators and pests that says, "I'm trouble."

The common wasp stings animals that threaten it. Animals that get stung learn to avoid wasps. Hornet moths, wasp beetles, and hoverflies don't sting. But they don't have to. They have markings that mimic the markings of the common wasp. Animals that

NATURE NOTE: *When chickadees find their nests threatened, they hiss. Scientists think the hissing scares nest-robbing animals by making them think there's a snake nearby.*

This fly can't sting, but it looks like it might

have learned to avoid wasps also avoid these harmless insects. Even though these insects are harmless, their ability to mimic the common wasp helps save them from becoming another animal's meal. A plant or animal that is not itself bothersome or dangerous but wards off predators by looking like something that is is called a **Batesian mimic**.

NATURE NOTE: *The mimic octopus, which is found in the Pacific Ocean, can change its color and shape to mimic other animals, including shrimp, jellyfish, and crabs.*

This harmless beetle mimics a dangerous wasp

TROUBLE TIMES TWO

Another kind of mimic is a **Müllerian mimic**. Müllerian mimics not only mimic a creature that is harmful or undesirable, they also are harmful or undesirable themselves. How does the mimicry help them? By increasing the likelihood that a predator will learn to leave creatures with that particular look alone.

NATURE NOTE: *Monarch-like viceroy butterflies were once considered an example of Batesian mimicry. But recently scientists discovered that viceroys taste bad, too.*

The monarch butterfly is poisonous if eaten

✸ For example, butterflies called viceroys look like monarch butterflies. Both monarchs and viceroys contain chemicals that make birds that try to eat them sick to their stomachs. If a bird tries to eat a monarch, it quickly learns to avoid orange butterflies with black markings. If it tries to eat a viceroy, it learns the same lesson. Because birds get the same message from both kinds of butterflies, they learn to avoid them a lot faster than they would if only monarchs or only viceroys tasted bad.

QUICK CHANGE

Some animals that use camouflage to protect themselves from predators or to gain access to prey can change their disguise. Flounders are fish that lie flat on the bottom of the ocean. Just as the appearance and texture of the ground varies, the bottom of the ocean can look different from one spot to the next. But that doesn't stop flounders from using camouflage to protect themselves. They just change their coloring to match the surface upon which they rest.

NATURE NOTE: *Flounders can change their color to match the sea bottom within hours of moving to a new spot.*

NATURE NOTE: *A flounder trying to camouflage itself will even try to match an elaborate pattern such as a checkerboard.*

A flounder can mimic various color patterns

❄ The snowshoe hare, which is found in forests in northern North America, has a brown coat that helps it blend in with dead leaves and other debris on the ground. But each fall, the hare trades its brown coat for a white one. If it weren't for this white coloration, the hare would be easily spotted by foxes and other foes against the bright snow of winter.

LURING INSECTS

Some plants use disguises to attract insects. When insects fly in to check them out, they help spread pollen from one plant to another, allowing the plants to make seeds.

☞ Orchids, a type of flower, are masters of this kind of mimicry. Species belonging to the **genus** *Ophyrus* have parts that look and smell like a female bee. Male bees fly in to try to mate with the "female bee." In the process, they pick up pollen to spread to other *Ophyrus* plants.

Many orchids use mimicry to lure insects

Another kind of orchid found in South America has a flower that dances in the wind like a flying male bee. When a real male bees sees it, he zooms in to attack what seems to be a trespasser in his territory. In the scuffle, he picks up pollen to spread to another plant. Other orchids mimic smelly things that insects like, such as animal droppings or dead animals.

This orchid mimics bees to spread its pollen

FATAL ATTRACTION

At the bottom of the Mississippi River, a little pink worm wiggles around. A fish swims up to it for a tasty bite. Then, suddenly, a pair of jaws snaps shut on it! The worm is not a worm, but part of the tongue of an alligator snapping turtle. By mimicking a worm, the turtle is able to lure fish right into its mouth.

NATURE NOTE: *Chameleons are perhaps the most famous color-changing animals. Scientists think this ability helps them to communicate as well as hide.*

Fish swim right into the mouth of this turtle

Some tropical fish also use this kind of foolery. Certain types of angler-fish have one fin that looks like a worm. When other fish come up to eat the "worm," the anglerfish sucks them into its mouth.

The anglerfish mimics a worm to catch fish

❊   The African flower mantis, a type of large insect, plays a similar trick on land. The mantis looks like a bright flower. When an insect flies up to find nectar, a sweet liquid produced by flowers, it becomes the mantis's meal instead.

This mantis looks as harmless as a flower petal

SCENTS AND LIGHTS

Deceptive odors can attract prey, too. The female bolas spider, found throughout much of North America, hangs a ball of sticky silk from a silky string. Then she gives off a scent that resembles that of a female moth. When male moths in the area catch the scent and fly in to find the female, the spider traps them by hitting them with the sticky ball.

The skunk cabbage, which grows in bogs and swamps, also uses odor to attract insects.

The soil in which this plant grows doesn't have much nitrogen, a nutrient that plants need to grow. But the skunk cabbage has another way of getting nitrogen. It gives off heat and a stinky smell, like that of rotten meat. Flies and other insects in search of a dead animal on which to lay eggs fly in and get trapped in a little pool of water at the base of the skunk cabbage. As the insects decay, the plant gets the nitrogen it needs from their dead bodies.

The rotten scent of skunk cabbage lures flies

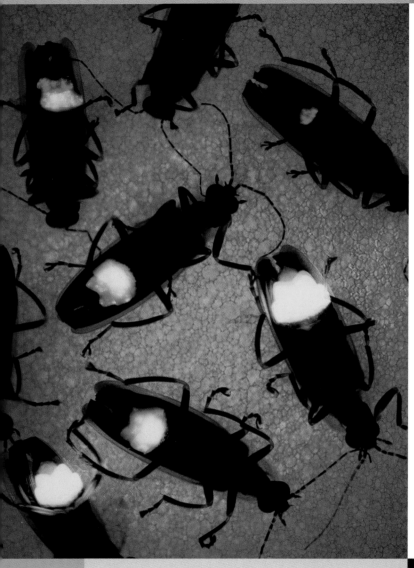

✎ *Photuris* fireflies mimic behavior to get their meals. Fireflies flash in the night to attract mates. Each species has its own pattern of flashing. Female *Photuris* fireflies mimic the flashing pattern of other species of the genus *Photinus*. The flashing attracts *Photinus* males. But instead of meeting a mate, they become a meal for the larger *Photuris* female.

NATURE NOTE: Photuris *fireflies can mimic the correct female flashing pattern response to the male flashes of several different species of* Photinus *fireflies.*

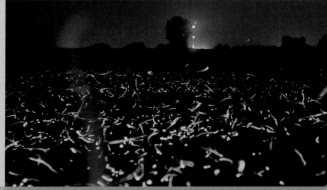

A firefly's flash can attract a mate or a meal

LIKE A GLOVE

From plants that smell like dead meat to bears and hares that look like snow, the world is full of creatures that use mimicry or camouflage to make their way through life. Like a hand in a glove, these creatures fit perfectly into the environment around them.

❧ The more we learn about living things, the more we realize how intricately each is woven together with others and with the environment in which they are adapted to live. When

A plant-eating weevil blends into leaves

we pollute air and water, destroy habitat, introduce foreign species, or do other things that directly affect some plants or animals, we often also indirectly affect many others. By considering the impact our actions have on the environment and its wild creatures, we can help ensure the future health and beauty of this amazing world, this world of wonder.

NATURE NOTE: *A certain kind of nudibranch (sea slug) takes on the color of coral (an ocean animal on which it lives) by eating some of the coral.*

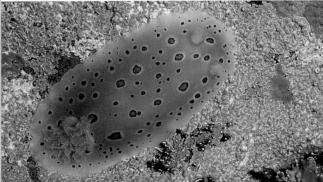

The horned lizard disappears into desert stones

Camouflaged stink bugs are hard to spot

WORDS TO KNOW

Adaptations *are characteristics that contribute to a living thing's ability to survive or reproduce.*

A **Batesian mimic** *is a harmless organism that mimics a dangerous or bothersome one.*

A creature that is **camouflaged** *has coloration that helps it to resemble its surroundings in a way that makes it hard to see.*

A **genus** *is a category of living things that is broader than species but narrower than family. A genus may contain many different species.*

Mimic *means to resemble another organism (as a noun, mimic is the organism that does the mimicking).*

A harmful or bothersome organism that mimics another harmful or bothersome organism is called a **Müllerian mimic***.*

Predators *are animals that eat other animals.*

Animals that are eaten by predators are called **prey***.*

INDEX